Edward Huntingford

The Parable of the Lock and Key

ISBN/EAN: 9783744761963

Printed in Europe, USA, Canada, Australia, Japan

Cover: Foto ©berggeist007 / pixelio.de

More available books at **www.hansebooks.com**

Edward Huntingford

The Parable of the Lock and Key

THE PARABLE

OF

THE LOCK AND KEY.

BY THE

REV. EDWARD HUNTINGFORD, D.C.L.

VICAR OF ST. SAVIOUR'S, VALLEY END, CHOBHAM,

*Late Fellow of New College, Oxford, and Head Master of Eagle
House School, Wimbledon. Author of "The Apocalypse
with a Commentary," and other Works.*

"To him give all the prophets witness."—*Acts* x, 43.

LONDON:

BICKERS & SON, 1, LEICESTER SQUARE, W.C.

---

*Price of cheap Edition 2d. for a single copy, and 10s. a hundred for distribution
by the Clergy, Religious Societies, &c.*

DRYDEN PRESS:
J. DAVY & SONS, 137, LONG ACRE, LONDON, W.C.

PREFACE.

THERE is a tendency just now to undervalue the importance of fulfilled prophecies as evidences of the truth of Holy Scripture. The following pages are written to counteract this. The writer has endeavoured, by means of a familiar illustration, to explain what seems to have been the Divine plan of Old Testament Prophecy; that its aim was to create in the minds of men the expectation of a Great Saviour, and at the same time to point out many very distinctive but very opposite characteristics by which He might be recognized when He appeared. Our Lord Himself declares that this was the object of Prophecy. "Search the Scriptures; for in them ye think ye have eternal life: and they are they which testify of me." (St. John, v. 39). When we read the Old Testament and compare it with the narratives of Christ's life, death, and resurrection, we observe an exact fulfilment in Him, and in Him alone of those born of woman, of a vast number of real predictions uttered by many different persons, in many different ages of the world, under many very different circumstances, and in many different ways; while some of the plainest of these are so contradictory to others equally plain that it must have been impossible to have imagined beforehand how they could ever be fulfilled in any one Person.

There are also many Christians who attach great importance to the general argument from prophecy, and yet find it difficult to believe in the reality of the prediction in any detail of future historical events. The following illustration is intended to meet this difficulty; and to direct attention to the fact that such predictions came not from the Prophets' own conceptions or anticipations, but from the suggesting or directing Spirit of the Omniscient. "No prophecy ever came by the will of man: but men spake from God, being moved by the Holy Ghost." (2 Pet. i. 21.)

Many of the passages quoted are taken from the family Bible of the Jews; not as being nearer to the Hebrew, but as giving the translation of the opponents of Christianity.

VALLEY END, CHOBHAM,
January, 1884.

The Parable of the Lock and Key.

WE have the highest authority for teaching by parables; and few things bring an argument more home to the mind than a well chosen parable.

Now there was a certain man who wished to make a lock which it should be quite impossible to open with any other than the right key. He was also particularly anxious that the mystery of his lock should not become known, until he had produced the key. But he could not make the whole with his own hands. He was obliged to employ workmen. How then was he to keep his secret? If he gave a plan of the whole to his workmen, and if they, working together with a full knowledge of what they were about, made all the wards and also put the lock together, they would be able to make a key to fit them. And then they might make any number of keys, or they might betray their master's secret, and so render the lock useless for its intended purpose.

He avoided this difficulty by employing several different workmen in different parts of the country. Having several factories he gave a drawing or a model of some single ward to some single workman in each factory, and told him to execute it in brass or iron. When the wards were all sent home he put them together with his own hands according to the plan which he had before designed, adjusting them so skilfully that they exactly corresponded with the key which he had prepared. The result was that only his own key could open the lock.

He then placed the lock in his window, and challenged the world to produce a key which would open it.

Many were the attempts made, but none succeeded. Picks and skeleton keys were tried, but none would answer; and many even doubted whether it was a lock at all. The secret was well kept. But when the time had fully come for the mystery to be revealed, the master himself produced the key. It looked just like any other key, but it was the true key: the key long ago designed and prepared; the key for which the lock was made. It was applied to the lock, and the lock was opened.

Such is our Parable, now for the interpretation.

He who designs the whole plan is God. The key is Christ. The lock is Prophecy. The individual wards are individual types or predictions. The workmen who made the wards to pattern are the Prophets.

Each Prophet in his own age and in his own country contributed his share towards the construction of the great enigma of Prophecy, to be eventually solved by the Appearance of Christ, and so to bear witness of Him. " The testimony of Jesus is the spirit of prophecy." " We have found him, of whom Moses in the law, and the prophets did write, Jesus of Nazareth." (S. John i. 45.)

The Prophets, therefore, were simply the workmen employed in forming the separate wards of the lock of prophecy. The model given them was either a vision impressed upon their imagination in sleep or in a trance, or some person, or the actions or circumstances of some person before them at the time, and not seldom some circumstances connected with themselves, their children or their country. This was, as it were, the drawing or the model in wood or clay, which they had to copy in brass or iron. In plainer words the enthusiasm of inspiration, the overwhelming and irresistible impulse of the Spirit of God, constrained them to describe the type or pattern before them in language which would often have been regarded by their contemporaries as profane and blasphemous, had it not been well understood that they were commisioned prophets, and that their words were intended to point out some characteristic of the Great Expected Messiah, the future deliverer of His people.

Indeed, the inspired language of the Prophets was not seldom derided by their profane contemporaries who regarded them as madmen. "Who hath believed our report?" says Isaiah, when describing the future sufferer who was to make atonement for His people. And the complaint of Ezekiel is to the same effect: "Then said I, Ah Lord God! they say of me, Doth he not speak parables?" (Ezek. xx. 49.)

They did not prophesy, therefore, out of their own minds, or from their own foreknowledge, but because they spoke under the irresistible influence of the Spirit of God. They confess themselves distinctly conscious of this. They declare that they are not speaking in their own name. The Prediction is introduced by the words "Thus saith the Lord," or some similar phrase. They do not pretend to know the full meaning of their words. In some cases they distinctly disclaim any knowledge whatever of the meaning of the vision they describe. " I heard, but I understood not : then said I, O my Lord, what shall be the end of these things ? and He said, Go thy way, Daniel : for the words are closed up and sealed (locked) till the time of the end." (Dan. xii. 8, 9.) "No prophecy of Scripture is of private interpretation (*i. e.* as the next verse shows 'is to be regarded as coming from the prophet's own unaided mind'). For no prophecy ever came by the will of man : but men spake from God, being moved (*literally* 'borne along') by the Holy Spirit." (II Pet. i. 20, 21.)

The prophets, therefore, were often perplexed in their own minds at the strangeness and apparent contradictions of their prophetic utterances. They could not conjecture how any key could ever be found to open a lock containing wards of such very opposite shapes. Nay, the Apostles represent not only the prophets, but even the Angels of heaven as studying with wondering interest the descriptions of the Coming Saviour. "Of which Salvation the Prophets have enquired, and searched diligently, who prophesied of the grace that should come unto you, searching what, or what manner of time the Spirit of Christ which was in them did signify, when it testified beforehand the sufferings of Christ, and the glory that should follow . . . . . which things the Angels desire to look into." (I Pet. i. 10-12.)

The workmen, then, employed by God to make the separate parts of this wonderful lock were the prophets; and of these Moses was the first. Let the unbeliever take any view he pleases of the story of Eden. Let him call it a tradition handed down to Moses, let him call it a myth or an allegory invented by Moses, or whatever else he pleases : the fact is undeniable that this ancient story contains the three primary wards of the great lock of prophecy.

It is a most distinct, though figurative prediction of an event to take place at some future time. It represents man as tempted and deceived by a power external to himself ; and it predicts that one born of a woman would at some future time destroy the power of this Tempter.

These, then, are the three primary wards of the lock of prophecy, and the true key must combine in Himself that which corresponds to each of them.

1. The true Champion of Man must be born of a woman; must be one of the descendants of Eve.

2. He must destroy him who has been the cause of man's sin and death ; He must crush the head of the Tempter.

3. He must Himself suffer; He must receive a painful but not a fatal wound in the process. He must be bitten in the heel while in the act of crushing the Serpent's head.

" I will put enmity between thee and the woman, and between thy seed and her seed ; it shall bruise thy head, and thou shalt bruise his heel." (Gen. iii. 15.)*

If subsequent workmen had added nothing to these

---

* It is of some importance to observe here that the word rendered "it" in the above passage is the common Hebrew three primary wards of the lock of prophecy, the history of the whole human race could not have shown any other key except Christ capable of opening it. No other child of woman has ever so effectually resisted the power of the Tempter, or suffered more in the process. No other man has shown so much Divine power over evil in His own person or imparted so much spiritual strength to His Disciples. No other has ever even claimed to be the Destroyer of the Evil One. No other has crushed the Serpent's head and been at the same time bitten by him in the heel.

But these wards have been elaborated in a most remarkable manner by subsequent workmen. We will pass on to another: "God so loved the world, that he gave his only begotten Son, that whosoever believeth in him should not perish, but have everlasting life." (St. John iii. 16.) Christ was to die, not, as the scoffing infidel misrepresents the matter, "to satisfy the anger of an infuriated Despot in heaven;" but as the evidence of the extreme hatefulness and misery of sin and the· inconceivable love of the Father for the sinner. Who was to prefigure this profound truth? What was to be the model from which this ward of the lock was to be

---

pronoun of the third person singular, commonly translated "He." So that we may fairly translate the passage, "He shall bruise thy head." The word "seed" may also mean either the human race or one special child of the woman. The context only can determine in what sense the word "seed" is to be taken in any passage where it occurs. Thus Gen. iv, 25, it is used of one individual, Seth ; "God hath appointed me another seed instead of Abel."

made? One whose name was High Father; and whose Son should submit to die in obedience to the command of his Father.

The name Abram means "High Father." What a significant name! Take any view you please of the book of Genesis or of the narrative of this most remarkable transaction; there stands the fact, the written narrative, the significant name, the account of this extraordinary strain, so to speak, upon filial obedience and paternal love. Why the name itself is suggestive and typical of the "Heaven Father" of which name Max Müller makes so much as appearing in the religions of all the most ancient races as the earliest name of God.

It is almost literally the Father of Heaven, for it means the "Father of Altitude," the "Father of the height" a Hebraïsm for the "High Father." It was changed afterwards, purposely that it might be significant of another great truth, that, although God designed to make the seed of Abram, through Isaac, a chosen people, He did not do so as a respecter of persons, but because He purposed to bring a blessing upon all nations by means of this chosen seed. And so his name was changed from "High Father," to Abraham, Father of a multitude of nations. "Neither shall thy name any more be called Abram, but thy name shall be Abraham, for a father of a multitude of nations have I made thee." (Gen. xvii. 5.)

And to point out still more plainly the Divine purpose of all this; to make this ward of the lock appear at the time still more intricate and yet to suit more exactly the true key when produced, these words of promise are added. "And all the nations of the earth shall bless themselves by thy seed." (Gen. xxii. 18.) These words occur with slight variations five times in the book of Genesis, and should be compared with two other passages, one in Isaiah and the other in the Psalms, each manifestly referring to the future Messiah. The words are the same as those of Genesis. "He that blesseth himself in the land, shall bless himself by the true God." (Isai. lxv. 16.) "His name shall endure for ever . . . . by him shall all nations bless themselves." (Ps. lxxii. 17.)

Here is another ward of the lock with something added on to make it still more difficult for any key except the right one to pass over it. The future Champion of Man is to be the only and well-beloved Son of the High Father; He is to submit to be sacrificed in obedience to that Father; and yet He is to be One Whom all nations are to worship, in Whom all the families of the earth are to bless themselves. And in Isaac's deliverance from this willing death the Apostle recognizes and has a right to recognize a dim foreshadowing of the Resurrection of Christ. "By faith Abraham, when he was tried, offered up Isaac: and he that had received the promises offered up his only begotten Son. Of whom it was said, That in Isaac shall thy seed be called: accounting that God was able to raise him up, even from the dead: from whence also he received him in a figure (*literally* in or by a parable.)" (Heb. xi. 17-19.) That is to say the act of the substitution of the ram for Isaac was when narrated a Parable indicating both the Resurrection of Christ and of all who die in Christ, and the substitution of a Victim for the whole seed of Abraham.

We may regard this, I think, as a very important addition to the third of those wards of the lock mentioned at the beginning; the suffering of Man's Champion; the wounding of the heel. And, as we cannot describe all the wards of this wonderfully intricate lock without giving an analysis of the contents of the whole Bible, it will be better to confine ourselves to the lines laid down for us in the story of Eden.

It will be sufficient for our present purpose, and will give us a keener appreciation of this wonderful work, if we trace out these three leading ideas, if we observe the gradual perfecting by the hands of separate workmen of these three principal wards of the lock of prophecy; that the Future Champion is to be—1. Born of Woman. 2. The Conqueror of the Tempter. 3. One wounded in the battle with him.

In the seemingly inhuman command given to Abraham, his anguish of soul, his ready obedience, and still more in the mental agony but willing submission of Isaac, we have an enigma which demands solution, a very secret detail added to that portion of the lock which requires a suffering Champion as the only key to open it.

To this we may add the institution of Sacrifice, the Divine but very strange command so often repeated to shed innocent blood to cover the guilt of those who have listened to the Tempter, and so broken the Divine law; the principle laid down apparently from the beginning, but if not from the beginning, certainly from the time of Moses, that "without shedding of blood is no remission " of sins. (Heb. ix. 22.)

In connection with this, then, we pass on to another most remarkable story.

The people of God are in the wilderness. They murmur and sin against God. He punishes them by sending among them a number of venomous serpents. Then Moses, at God's command—the same God, be it observed, who had so strictly forbidden the making of any molten image, or attaching any kind of Divine influence to it; who had so repeatedly condemned any imitation of the Idolatries of Egypt— sets up on a pole a Serpent of brass, and at once the poison of the serpents is harmless to all who look upon the brazen image on the pole. What a very extraordinary story! We Christians believe that it is a true story, and that it happened just as Moses describes it. But if the unbeliever denies this—What does he gain by the denial? He cannot deny that there is the story, whether it be a true narrative, or a tradition, or an allegory, in any case there it is written down in that book which professes to come from God and to predict long beforehand the nature, work, and suffering of the future Champion of man. Here then we have again the Serpent, the wounding of the heel, the crushing of the head; the venom of the Serpent rendered harmless by the lifting up before the eyes of the serpent-bitten people of One made in the likeness of the Serpent. What a mystery is here? The Israelites thought a great deal of this brazen figure, and no wonder. They preserved it for many centuries. They regarded it as a sacred relic. They made a sort of fetish of it. So much so that the pious King Hezekiah wisely determined to destroy it. He did so because the people were foolish, not because he doubted the story or failed to recognize that it was a mysterious sign awaiting solution.

They were right in regarding it as a mysterious sign, an enigma awaiting solution in the future. They doubtless saw that the transaction would have been childish in the extreme, and the miracle most idle and unmeaning, were not this brass copy of the Serpent lifted up on high a mystery, a symbol veiling some great truth not yet made known to them. Modern experience proves how easily an object associated with some great event becomes a fetish, and then the sooner it is destroyed the better. But this deep veneration for a relic proves that those who venerate it believe the story of its origin and the fact that it represents a mysterious truth.

A piece of wood, supposed to have been part of the true Cross, would not be worshipped, if men did not believe the narrative of the Crucifixion. Pilgrims would not flock in thousands to see the Holy Coat if they did not profoundly believe in Him who, as they are told, had worn it. Neither would the Nations of Europe have gone in a body to rescue the Holy Sepulchre out of the hands of infidels had they not profoundly believed in the mystery of Redemption and in the story of Him, who died upon the Cross, was buried, and rose again from the dead.

The fact, therefore, that this Serpent was so ong

preserved and so deeply venerated adds considerably to the importance of the story connected with it as a ward, and an important ward, in the lock of prophecy. Alone it would not count for much. A single ward never counts for much in a lock. Any key, or, at any rate, many keys will open a lock that contains only one or two simple wards ; but only one key will answer when the wards are many, intricate, contradictory, and skilfully combined.

As Christians we have the highest authority for attaching importance to this extraordinary transaction, for Our Saviour appeals to it as a symbol significant of His Crucifixion. "As Moses lifted up the serpent in the wilderness : even so must the Son of Man be lifted up : that whosoever believeth in Him should not perish, but have eternal life." (St. John iii. 14, 15.)

But why, it may be asked, should the form of a serpent have been chosen as the symbol of man's future Saviour, born of a woman, wounded by the Serpent, but victorious over him ? We answer. Because it symbolizes two most fundamental articles of our Christian faith, two truths connected with the nature and work of Christ, so apparently contradictory to one another, that no creature could possibly have imagined beforehand how they could be both realized in the person of one Saviour. What figure could better represent at one and the same time the deep humiliation and the triumphant victory of the Great Conqueror of the Serpent, and especially of this victory gained by the act of humiliation ! He took not on Himself the nature of a spotless angel or of unfallen Adam, but "sinful flesh." He was born of a woman; of a pure virgin, indeed, yet one who like all before her, was tainted with the blood poison of the Serpent, and herself needed a Saviour. He was sent "in the likeness of sinful flesh." (Rom. viii. 3.) Contact with the Divine indeed cleansed the human; just as the touch of Christ healed the leper without defiling the Healer. But by His Incarnation, and Birth, and especially by His Crucifixion, the Son of God willingly subjected Himself to the full burthen and curse resulting from the original venom of the Serpent, and by that act of submission conquered the Serpent, and held him up in triumph to the eye of faith.

Both these aspects of the Cross are made of great importance by the Apostles. Jesus on the Cross is held up to the eye of faith as He who was the cursed one, and also as He who conquered the cursed one; the Serpent, and the Serpent's Conqueror. What a mystery of contradiction we have here! What a strangely intricate ward of the lock of prophecy was made by the workman Moses when at God's command he fashioned and set up that brazen Serpent, and wrote down the description of the healing of the serpent-bitten people!

Here then is the description of that part of the true key which exactly fits this very perplexing part of the lock. "As Moses lifted up the serpent in the wilderness, even so must the Son of Man be lifted up: that whosoever believeth in him should not perish, but have eternal life." (St. John iii, 14, 15). "He hath made him to be sin for us, who knew no sin; that we might

be made the righteousness of God in him." (2 Cor. v, 21).
"Christ hath redeemed us from the curse of the law,
being made a curse for us: for it is written, "Cursed of
God is every one that hangeth on a tree." (Gal. iii, 13.)
But in taking the likeness of the cursed one He trium-
phed over him, and held him up as the trophy of victory.
"My God! my God! why hast thou forsaken me?" were
the words of Him who made Himself accursed for us.
The loud cry out of the darkness, which shook the
earth to its centre, and wrung the words of conviction
from the lips of the Roman soldier, was the triumphant
shout of victory. "It is finished." The heel has been
sorely wounded, but the head of the Great Enemy has
been crushed. Henceforth Jesus, once hanging upon the
cross, is the Healer of His serpent-bitten brethren, the
great Champion who has conquered by suffering, who has
overcome death by death, evil by good; and who calls
upon all men to walk in His steps to victory, to take up
His cross and follow Him. For He has "blotted out
the hand-writing of ordinances that was against us,
which was contrary to us, and took it out of the way,
nailing it to His cross; and having spoiled principali-
ties and powers, he made a show of them openly,
triumphing over them in it." (*i. e.* in His cross) (Col. ii,
14, 15.)

We cannot imagine that Moses, when he was making
to pattern this most remarkable ward of the lock,
when he was describing this strange miracle, had any
conception of the corresponding part of the key which
was to answer to it; there is no reason for believing that

he attached any prophetic meaning at all to it. He simply did what God had commanded, and described what he had done. Just as individual workmen in a great factory make their own appointed portion of some elaborate machine, without any knowledge whatever of the inventor's whole plan.

The close connexion of shame and victory symbolized by this story of the Serpent on a pole, is dwelt upon frequently by the Psalmist. The words of the twenty-second Psalm seem made, as it were, to express the anguish of the Sufferer on the Cross, whatever may have been the model, if any, immediately before the eyes of the writer. He is forsaken by God and yet is full of trust in Him. He is derided by all who see Him on the cross as one cursed of God as an impostor. And yet He declares that He is not an impostor, and predicts that the result of His rejection by God and man will be the salvation of the world. " I am a worm and no man : a reproach of men, and scorned by the people. All they that see me laugh me to scorn ; they distort the lips, they nod the head, saying, Trust on the Eternal, let him rescue him ; let him deliver him, seeing he delighted in him. . . . . They gaped upon me with their mouths, as a tearing and roaring lion. I am poured out like water, and all my bones are out of joint. . . . Dogs have compassed me : an assembly of wicked ones beset me round ; like the lion, my hands and my feet. I may tell all my bones ; behold, they look and stare upon me. They divide my garments among them and cast lots upon my vesture."

The foregoing is the rendering of the Psalm in the Jews' family Bible. That translation, therefore, must be above suspicion; and it proves the faithfulness of the Jews in the preservation of the Hebrew text of their Scriptures. The reader will observe that this literal translation of verse 16 does not materially lessen the significance of the prophecy, as pointing to Jesus on the Cross. The sentence "As a lion, my hands and my feet," requires the insertion of a verb supplied by the context in order to make sense. The words evidently mean " as a lion (they rend) my hands and my feet." Our translation corresponds with the Septuagint; and it is impossible to determine whether the translators of the Septuagint had before them a text differing from the present Hebrew; or whether they omitted the words " as a lion " and gave what they conceived to be the meaning of the passage. Take the passage as we will, we have in this Psalm the addition of an important detail to that ward of the lock of prophecy constructed in the beginning : " Thou shalt bruise his heel." This Psalm certainly holds up to the eye of faith a Mighty Sufferer, forsaken by God and despised by men; yet One who is confident of final victory, and sees before Him the salvation of the whole world as the result of His suffering. " I will declare thy name unto my brethren ; in the midst of the congregation will I praise Thee. . . . . All the ends of the world shall remember and turn unto the Lord ; and all the kindreds of the nations shall worship before Thee."

We must not dwell much longer on the construction of this ward, " Thou shalt bruise his heel." It was completed by later prophets, especially by Isaiah and Daniel.

It is impossible to determine whether Isaiah had any type or pattern before him, when he described under the enthusiasm of inspiration the Servant of Jehovah, who was to " be exalted, and extolled, and be very high : " and yet whose " visage was marred more than any man, and his form more than the sons of men." (Isai. lii, 13, 14). It must have been a strange model, if he had any. And at any rate his description fully answers to One, and only One Person in the history of the whole world,—a Conqueror, victorious after being rejected by God and man ; and One by whose stripes and death His people were pardoned for their sins. We search in vain for such a Conqueror in all History Ancient and Modern. We find such a One described in the simple writings of some Galilæan fishermen ; and nowhere else. Of what other could the Prophet say these words ? " He hath borne our griefs, and carried our sorrows : yet we did esteem him stricken, smitten of God, and afflicted. . . . . All we like sheep have gone astray ; we have turned every one to his own way ; and the Lord hath laid on him the iniquity of us all." Yet it is He who is to make men righteous, and to be rewarded as a Conqueror : " By his knowledge shall my righteous servant justify many ; for he shall bear their iniquities. Therefore will I divide Him a portion with the great, and He shall divide the spoil with the strong." (Isai. liii, 3-12.)

Daniel certainly had something to copy. He faith-

fully described the patterns before him, but acknow-
ledged, as we have observed, that he did not understand
the meaning of his miraculous visions.

We believe that Daniel lived during and just after
the Babylonish Captivity, and that he saw and described
his visions at that time.   His drawings, from which he
was to finish the wards of the lock assigned to him,
were certain miraculous and allegorical visions which
he saw in his sleep, or in a trance, and described afterwards.   And during the exhibition of these visions he
saw angels, and asked them questions and heard their
answers.   These also he faithfully recorded, though he
confessed that he could not understand them.

There was an obvious reason why this workman should
have patterns of a very peculiar character to work from.
He was to give, as it were, the finishing touches, and to
add very delicate details to several of the most intricate
wards of this wonderful lock.   And so, instead of a type
or pattern, he had, as it were, some very elaborate and
carefully finished drawings given him.   These were impressed upon his imagination during sleep, or when he
was entranced, and a few words of explanation were
occasionally added.   The result was the composition of
a number of symbolical pictures of the future of God's
people, and of the kingdoms connected with them up to
the very end of the Christian dispensation and the
second Advent of Christ, which have been verified by
history so far as it has yet gone, and are being verified
more and more clearly in every succeeding age of the
world.

It is only natural that the thieves and robbers of every age who want to prove either that prophecy is no lock, or one which can easily be picked by a false key, should hate this workman who gives them so much trouble. If Daniel lived and wrote in the sixth century before Christ, unbelievers must admit the reality of prediction, and so the very strongest evidence of the truth of the Bible. They have therefore left no stone unturned in their efforts to prove that Daniel's prophecies were after-prophecies or impostures, written at the time of Antiochus Epiphanes, about 170 years before Christ. Their arguments have been completely refuted by that learned Hebraïst and profound theologian who has so lately gone to his rest. But after all what do they gain by calling Daniel an impostor and accusing him of falsehood when he gives the exact date of his writings in his own book? Their argument is a very simple one. There cannot be, they say, such a thing as a real prediction in any detail of historical events in the remote future. But if we admit that Daniel saw and described his visions at the time when he distinctly tells us he did so ; then, since his prophecies about Antiochus Epiphanes were certainly fulfilled, we must admit that he was a true prophet, that he was an inspired writer, that the Bible is true, that Christianity is true, and that we incur grave responsibility in denying its truth. Rather than do this we will beg the whole question and assume at once that the close correspondence which can be observed between some of Daniel's visions and the events preceding the Age of Antiochus proves that he lived at that time and described them in the allegorical language of pretended visions. Such is really their argument, and, as Dr. Pusey clearly shows, their unbelief precedes their criticism, and is not its result. They disbelieve the fact of real prediction first, and then criticise the book to support their unbelief.

But what do they gain by it? Literally nothing. Nothing suffers except the character of Daniel as a speaker of the truth when he gives the date of his own writings. These critics admit quite enough for the Christian's purpose. They admit that Daniel's pro-phecies were written at least 170 years before Christ, and that many of them were intended by the writer to be predictions of events long subsequent to the age of Antiochus, extending as they do to the overthrow of all earthly kingdoms by the kingdom of One, whom he calls the Son of Man; to the Day of Judgment, and of the resurrection of the dead.

We Christians believe that Daniel was a true Prophet. And it is remarkable that Jesus Himself calls him a prophet, and speaks of some of his prophecies as awaiting fulfilment in the future (S. Matt. xxiv, 15). We also believe that he was a truth-speaking man, and that, therefore, what he distinctly tells us about the date of his writings must be true. But for the sake of argument we may grant that he may have written, or rather that someone assuming his name may have written, what we call the book of Daniel 170 years before Christ, instead of about B.C. 540.

This is quite enough for our present purpose, for

Daniel speaks of "an Anointed One" who is "to be cut off with none to help him." (We give the Jews' translation of Daniel in their English family Bible). The unbeliever may say that this refers to some anointed priest or king known to the writer although unknown to us. But we answer, Daniel does not merely say that an Anointed One is to be cut off. He foretells that this is to happen at an appointed time, and at a very remarkable time. Two most contradictory things, he says, are to happen simultaneously, or nearly so. The greatest of blessings is to be conferred upon God's people and City and Sanctuary ; and yet this Anointed Leader is to be cut off with none to help Him, and then a Leader is to come in with a mighty people like a flood, and to destroy both City and Sanctuary, and to render them desolate for ever. How could both these things happen at the same time ?

The answer of the Christian is simple and satisfactory. Jesus is proved to be the true key by the facility with which He opens the lock. The prophecy contained in Dan. ix, 24-27, is hopelessly contradictory until it is applied to Christ, and then it is almost like a contemporary narrative, far closer to history than anything Daniel says about events before or at the time of Antiochus. It is a very providential circumstance that the Jews, the open enemies of Christ, have preserved this prophecy and borne witness to its composition ages before Christ came into the world. It is no less providential that learned unbelievers and half-believing Christians of the present day are unanimous in affirming that it cannot possibly have been written later than about 170 years before Christ.

This is certainly most providential, for, if it were not the case, these words of Daniel would certainly be regarded as the clearest proof that the writer was a Christian impostor. No prophecies of Daniel about the events of the age of Antiochus are anything like so close to history as these words are to the doctrines of Christianity, and the judgments which fell upon the Jews for cutting off their Messiah.

The Jews faithfully preserving the Hebrew text have yet done their best, in their English translation, to make the application of these words to Christian times less obvious; and yet the following, which is their own translation, is quite sufficient for the Christian argument. The words in brackets express the legitimate sense of the Hebrew.

" Seventy weeks (periods of seven) are determined upon thy people and upon the city of thy holiness (Hebraism for thy holy city), to restrain the transgression and to make an end of sins, and to make an expiation for iniquity, and to bring in perpetuity righteousness (eternal righteousness), and to seal up vision and prophecy, and to anoint the holy of holies." (This expression generally means the most holy place, but Chron. xx. 3, 13, it is used of Aaron, as translated by the Jews, " Aaron was separated to sanctify him as most holy," *literally* " holy of holies.") " Know therefore and understand, that from the going forth of a sentence to restore and to build Jerusalem unto an Anointed One, a leader, shall be seven weeks; and in threescore and two weeks shall street and ditch be built again, even in troublous times." Here the reader will observe how the punctuation is altered, and the word " in " inserted before 62 weeks. The following passage shows that the 62 weeks terminate with the cutting off of the Anointed Leader, and not with the completion of the city, which is to coincide with the end of first period of seven weeks. The whole period given is 70 weeks. This is then subdivided into three periods of 7, 62 and 1.

Three events are also mentioned in the same order: the completion of the city, the cutting off of Messiah, and the final desolation of the Jews' city and sanctuary by a leader and people who should come upon them like a flood. The first period of seven clearly refers to the first event, the completion of the street and ditch of the city, which had been destroyed by the Babylonians, and this dates from a decree of the Persian king, authorizing the rebuilding of the city; then 62 weeks elapse and the Anointed Prince is cut off; then the covenant is confirmed with many in the last single week of the seventy, in the midst of which sacrifice and oblation are put an end to; then up to the consummation war and desolation are to be poured upon the desolate. The Jews' translation of the next verse shows that this cutting off of the Anointed One occurs at the end of the 62 weeks, not of the 7 weeks. " And after the three score and two weeks shall an Anointed One be cut off, and there is none to help him" (*literally* none with him). " And a people of the coming leader " (*literally* of *a* leader who shall come) " shall destroy the city and the sanctuary; and his end " (or the end of it, *i.e.* of the sanctuary) " shall be with a flood, and unto the end of the war desolations are determined. And he shall make a powerful covenant with many for one week : and half a week " (or in the midst of the week) " he shall cause sacrifice and oblation to cease ; and upon the skirt of the abominations " (or to the extremity of abomination) " there shall be a desolator, even until the consummation, and that determined shall be poured upon the desolate " (*i.e.* upon the Apostate Jews).

Now let us carefully observe what is stated in this chapter, even according to the Jews' rendering.

The captivity of Judah in Babylon was to last 70 years. Daniel knew that the predicted time of restoration was drawing near. In common with the other prophets he anticipated a full and final restoration of Israel's glory at that time under their predicted king of the line of David. But the actual return of the exiles, and the literal rebuilding of the city and temple at the end of 70 years was to be only the beginning and the earnest of the final restoration of his people and the glory of the Great Son of David. Much was to be done and suffered before that time. While Daniel, therefore, is praying for the literal restoration of Israel, a miraculous vision is revealed to him, probably while he is in a trance. The Angel Gabriel comes to explain the matter to him. The 70 years, he shows him, refer only to the literal rebuilding of the city and temple. They must be multiplied by seven before the Great David comes. There must elapse not 70 years but 70 sevens of years. And when He comes, He will not come for any purpose such as men think; but he will come to make an expiation for iniquity, and to introduce a principle of everlasting righteousness and to be anointed as most Holy. And this Anointed One, instead of being a great earthly king as men expect, will " be cut off with none to help him." And this His cutting off will make the expiation for sin, and will put an end to all other sacrifice and oblation. And as far as concerns that literal city and temple, about which you think so much, know, says the Angel, that these shall be utterly destroyed after the Anointed One has been cut off, for a very different kind of leader and a mighty nation shall come down upon them with the suddenness and violence of a flood, and make them waste and desolate for ever.

Daniel, we know, professes not to understand his prophecies. These words of Gabriel, heard during his trance, must indeed have sorely perplexed him. Sin is expiated, everlasting righteousness introduced, the Most Holy to be anointed. And yet this Anointed Prince is to be cut off with none to help him, sacrifice and oblation are to be abolished, and the city and sanctuary are to be destroyed by war and to remain desolate until the end. With such sayings as these before them, well might our Lord say to the two Disciples at Emmaus : " O fools, and slow of heart to believe all that the prophets have spoken ; ought not Christ to have suffered these things (to have been cut off, with none to help him) and to enter into his glory?" (St. Luke xxiv. 25, 26.)

There are other minute details connected with the Crucifixion, foreseen by God and therefore added by His direction to that ward of the lock, which belongs to the wounding of the heel. Besides the rending of the hands and feet, the gall and vinegar, the casting lots upon the vesture, &c. of the Psalmists, St. John recognizes in the circumstance that the soldiers "brake not His legs" but pierced Him with a spear, the fulfilment of the words of Zechariah xii. 10, "They shall look unto me whom they pierced"; and the singular command given to the Israelites to kill the paschal Lamb without breaking any of its bones (St. John, xix. 32-37.)

Such, then, is the completion of that ward of the great lock of Prophecy, which was constructed, as it were, in the rough by Moses when he wrote the words, "Thou shalt bruise his heel."

The future Champion of man was to be given as an only Son in sacrifice by the High Father, but rescued from death. He was to be lifted up before His people as one cursed of God. He was to be scourged, despised and rejected by God and man, yet by His stripes man was to be healed of sin. He was to be cut off with none to help Him, when there was to be made an expiation for sin, when sacrifice was to be abolished, and the holy city and temple destroyed and made desolate for ever. Not a bone of Him was to be broken, but His hands and feet were to be torn; He was to be stabbed with a spear; He was to be gazed upon, mocked and derided; lots were to be cast upon His vesture; vinegar was to be given Him to drink. These and many others were strange predictions, incomprehensibly mysterious utterances, intricate details of the ward, " Thou shalt bruise his heel." Moreover the very title over His cross proclaimed Him the promised Saviour and King of the Jews, the predicted Scion of David, the Branch which should grow out of the root of Jesse. (Isai. xi. 1, Jer. xxiii. 5, xxxiii. 15, Zech. iii. 8, vi. 12, S. Matt. ii. 23.) As written in Hebrew this title would be, " Joshua the Branch, the King of the Jews." No wonder that the Pharisees did not like it. They felt ill at ease when they read it. So much so that they begged Pilate to have it changed. But he was in no humour to gratify men who had persuaded him to commit such a judicial murder against the goads of conscience and his own convictions. " What I have written, I have written," was his sullen and peremptory answer. (See *Smith's Dictionary of the Bible*, under the word "Nazarene.")

Such, then, is the lock of Prophecy.

By no other key except Christ, even by the confession of unbelievers, has any one yet, with any appearance of success, attempted to open it. But Jesus on the cross solves all its most perplexing contradictions. " O fools and slow of heart to believe all that the prophets have spoken; ought not the Anointed to have suffered these things, and to enter into his glory ?"

Such, especially, is that ward of the lock given in the story of Eden, " Thou shalt bruise his heel," which requires, as the only key which can fit it, a Champion severely but not fatally wounded in His deadly encounter with man's Great Enemy.

But was Jesus man's deliverer ? Though like him born of a woman, was He unlike him in His power and in His work ? Though suffering like him by the wound of the Serpent, did He crush that Serpent beneath His wounded heel ?

It is admitted that Jesus was born of a woman, in the city of David, a scion or branch out of His roots. It is an historical fact that He was bruised in the heel, that " He was crucified under Pontius Pilate." But has He crushed the Serpent's head ? Has He proved Himself the true key by passing smoothly over this most important ward of the lock, and thereby opening the treasure house of everlasting life ? Has He conquered sin and death ? By the admission of His bitterest enemies He conquered the Tempter, at any rate in His own person. Alone of men He lived and died without sin. Alone of those born of woman He could say in the presence of those who hated Him, " Which of you convinceth me of sin ?" (S. John viii. 46.) We have the recorded sentence of the Roman Judge. " I find no fault in this man .... No, nor yet Herod." (S. Luke xxiii. 4, 15.) And we know that He was crucified not because He was proved guilty of any crime, but because He claimed to be the true Messiah, the Son of God, and the King of Heaven. Micah foretold that the Jews' Messiah should be born in Bethlehem, but should be an Eternal Being. The following is the Jewish translation of his words : " Thou Bethlehem Ephratah, destined to be little among the thousands of Judah, out of thee shalt he go forth unto me, to be a ruler in Israel ; whose goings forth (origin or descent) have been from of old, from the days of antiquity (*literally* Eternity) (Micah v. 2.) Daniel prophesied that He, though a Son of Man, should be an everlasting king. " I saw in the night visions, and, behold, one like a Son of Man came with the clouds of heaven, and came to the Ancient of Days, and they brought him near before him. And there was given him dominion, and glory, and a kingdom, that all people, nations and languages, should serve him : his dominion is an everlasting dominion, which shall not pass away, and his kingdom that which shall not be destroyed." (Dan. vii. 13, 14.) Such a King Jesus could not be unless He conquered death. It was therefore predicted of Him that He should overcome death for Himself and for His people by rising from the grave. " Thy dead men shall live, together with my dead body shall they arise. Awake and sing, ye that dwell in the dust." (Isai. xxvi. 19.) We believe that Jesus rose again from the dead, and thereby conquered death. We believe the word which He has spoken. " The hour is coming in the which all that are in the graves shall hear his voice, and shall come forth." (S. John v. 28, 29.) We believe that He is the King of heaven, and that " He must reign, till he hath put all enemies under his feet " . . . and that " the last enemy that shall be destroyed is death." (1 Cor. xv. 25, 26.)

The language of prophecy is distinct as to the Royalty and Divinity of man's great predicted Champion; that son of a woman who was to crush the powers of evil. "Unto us a child is born, unto us a son is given: and the government shall be upon his shoulder: and his name shall be called Wonderful, Counsellor, the Mighty God, the everlasting Father (Heb. Father of eternity, *i. e.* one who should be eternal), the Prince of peace." (Isai. ix. 6.)

He was to be a Divine King. As such He was to receive the worship of all men. That which is impious in earthly kings, deceived by their flatterers, was to be given as His due to the predicted Son of David. These are the words of Psalm lxxii., as translated in the family Bible of the Jews. " In his day shall the righteous flourish; and abundance of peace, until there be no moon. He shall have dominion also from sea to sea, and from the river unto the ends of the earth. Before him the tenants of deserts shall kneel down; and his enemies shall lick the dust. The kings of Tarshish and of maritime settlements shall bring gifts: the kings of Sheba and Seba shall offer presents. Yea, all kings shall prostrate themselves before him: all nations shall serve him. For he shall deliver the needy when he crieth for help; the poor also, and him that hath no helper."

The model here was probably Solomon. The enthusiasm of inspiration hurries on the prophet to describe the dignity of One " greater than Solomon."

He was also to be a priest as well as a king. What a strange confusion of offices usually kept so distinct in the Law of Moses! Zechariah prophesied at a time when there was no king in Israel; but when all were looking for the promised Scion of David. These are his words as translated by the Jews. "Thus saith the Eternal as follows, Behold a man, his name is Tsemach (Branch, *i. e.* the Branch of David) and he shall grow up out of his place, and he shall build the temple of the Eternal : even he shall build the temple of the Eternal ; and he shall receive majesty, and shall sit and rule upon his throne ; and he shall be a priest upon his throne." (Zech. vi. 12, 13.)

The model here was "Joshua, the son of Josedech, the high priest." He was indeed a priest, and he helped at the rebuilding of the literal temple. But he was not " The Branch," for he was not a son of David, nor even of David's tribe; he was not a king; he had no royalty, no throne on which to sit and rule. The workman had before him a base model, a mere man, a model of clay. But the spirit guided him in the execution of the work. He fashioned a ward for the lock of costly metal. His language suits only but completely the true Joshua, the true Josedech ; Jesus, the Lord our Righteousness.*

---

* Jesus is the Greek form of Joshua, and Josedech is Jehovah Zedech, or the Lord our righteousness, which became a Jewish name after the utterance of Jeremiah's prophecy. In Jer. xxiii. 6, " The Lord our righteousness" is written as the name "Josedech" in the Septuagint ; which proves that it was regarded as a name of the future Messiah by the Jews of that day, nearly 300 years before Christ.

No other son of a woman except Jesus of Nazareth, the village of branches, ever even claimed this Eternal Priesthood after the order of Melchizedech, the King of Righteousness; this heavenly throne; this universal and unending dominion; these attributes of Deity. But it is an historical fact that Jesus has been worshipped as Lord, obeyed as King, and approached as Priest for more than eighteen centuries.

He is King by lineal descent, as the Scion of David. King by divine appointment; though the kings and rulers of the earth take counsel together against Him. (Ps. ii. 2, 6.) King by right of conquest; going forth " conquering and to conquer." (Rev. vi. 2, xix. 16.) Above all He is a King, the pillars of whose throne are the love of His subjects and the free choice of millions of loving hearts.

No other founder of a religion has ever even claimed, much less received, such homage.

Can we count His subjects ? Those who have loved Him, adored Him, obeyed Him as their King, and worshipped Him as their God ? Those who have willingly suffered for Him the loss of all things; have loved Him to the end; and died for Him, exulting in the sacrifice ?

It is a simple fact which cannot be denied, that it was plainly predicted that the Champion of Man should be a King and Divine, the conqueror of sin and death; and also that He should win His throne by suffering.

Let the unbeliever take any view he pleases of Jesus of Nazareth, he cannot deny that He, who was crucified in the reign of Tiberius, has been a real King in all that constitutes royalty ever since, for He has been loved, obeyed and worshipped by millions of subjects. As such, therefore, He is the true Key, the only Key, which opened, without violence, the locked door of prophecy, and now throws wide open to all believers the treasure house of Eternal Life, the gates of the Kingdom of Heaven.

His work is not yet finished. Scripture tells us that it is not to be finished yet ; not until all man's enemies have been destroyed ; not until the number of the elect is filled up. Judah is not yet saved. Israel does not yet dwell safely. The heathen have not yet been all given to Him as His inheritance, nor the uttermost parts of the earth as his possession. (Ps. ii. 8.) But the mighty stream of conquest is daily gathering strength, in spite of all the opposition of the wicked and the unbelieving : and soon "the earth shall be full of the knowledge of the Lord, as the waters cover the sea." (Isai. xi. 9.)

Whatever they may think of Christ, none can deny, that His religion is the best of all religions, and has been and still is a conquering religion. According to the Scriptures it has been opposed and persecuted, but it has overcome all opposition. According to the Scriptures it has been fearfully corrupted ; but it has conquered in spite of corruptions. According to the Scriptures it is again in these latter days beginning to be rejected with contempt by the wise of this world, who allow themselves to be deceived by the old Deceiver, dressing up in modern form the old original falsehood, " Ye shall not surely die."

Meanwhile we have for our support the sure words of

prophecy: though " the kings of the earth set themselves, and the rulers take counsel together, against the Lord, and against his Christ, saying, 'Let us break their bands asunder, and cast away their cords from us.'" Nevertheless, " He that sitteth in heaven shall laugh, the Lord shall have them in derision," and then, when the time is come, " He shall speak unto them in his wrath, and vex them in his sore displeasure," and say, " Yet have I set my king upon my holy hill of Zion." (Ps. ii. 2-6.)

And so He, who, in the vision of Patmos, went forth at the first alone, sitting upon a white horse, " conquering and to conquer," shall come again " with clouds, and every eye shall see him, and they also which pierced him." (Rev. i. 7.)

When the warfare of Jerusalem is accomplished; when the dispersion of the Holy people shall be finished; (Dan. xii. 7.) " When the times of the Gentiles," during which " Jerusalem shall be trodden down" by them, " shall be fulfilled;" (St. Luke xxi. 24). When all the powers of evil shall be subdued; when Christ shall have " put all enemies under his feet;" (1 Cor. xv. 25). When the great Serpent, and the Wild-beast and the False Prophet have been defeated; then shall He come again, followed by all the armies of heaven, " riding upon white horses, clothed in fine linen, white and clean;" having on His head " many crowns," and " upon his vesture and on his thigh a name written KING OF KINGS AND LORD OF LORDS." (Rev. vi. 2, xix. 11-21.)